미용
영어

English for The Hairstylist

미용 영어

English for The Hairstylist

김세욱 김유정 김영주

구민사 Technology Computer
www.kuhminsa.co.kr

머리말

미용을 공부하는 학생으로서, 또 전문미용인으로서 영어라는 것을 반드시 공부해야만 하는 시대가 되었다.

과거에는 미용이라는 것은 기술적인 범주로 이론적인 기초 없이도 실기력만 있으면 가능하다고 생각했던 시기도 있었다.
그러나 세상이 발전하고 사람들의 의식수준이 발전하면서, 미용 또한 세계의 여러 가지 정보와 이론을 받아들이는 과정이 필요하게 되었다.

과거에는 어느 먼 나라의 특정 브랜드의 트렌드로 우리와 상관없는 것으로 생각되어지던 것이 수많은 정보와 실시간적인 교류가 가능해지게 됨에 따라서 우리의 생활 안에서 밀접하게 연관된 중요한 것이 되어버렸다. 무엇보다 이러한 정보의 실시간적인 교환이 가능해지면서 세계가 하나의 큰 고리로 연결되었다.

이제는 우리 미용현장에서도 수많은 외국인들을 만날 수 있으며, 미용인 또한 유학 및 다양한 목적을 가지고 해외로 나갈 일이 생기게 되었다.

또한 다양한 미용관련 원서나 잡지들을 접할 수 있으며 다양한 동영상 및 데이터들을 쉽게 구하고 접할 수 있게 되었다.

대부분의 우리는 외국에서 자라나고 영어를 자연스럽게 사용할 환경이 아니었으며, 미용인들이 영어를 전공한 학생처럼 많은 시간 영어를 공부하지도 못했을 것이기에 영어에 대한 느낌은 미용을 하는 사람에게 있어서 매우 높은 벽처럼 느껴지는 것이 사실이다.

그리고 지금까지는 피할 수 있으면 피해도 되고, 관심 있는 몇몇 사람만 조금 더 공부하면 되는 것으로 생각할 수도 있었다.

하지만 이제 더 이상 미용인에게 영어는 선택적인 대상이 아니며, 반드시 공부해야만 하는 것이다. 그 이유는 글로벌한 환경에서 영어를 이용하여 더 다양한 미용관련 지식 및 트렌드를 익힐 수 있으며, 우리 또한 더 많은 세계로 진출하여 일할 수 있는 기회가 많이 생겼기 때문이다.

이 책은 우선 헤어스타일리스트로서 미용관련 원서나 동영상 등을 통해 접한 미용관련 용어들을 익히고 이해하여 기본적인 지식을 습득 할 수 있도록 도와주고, 두번째는 현장에서 외국인 고객을 접대하면서 겪게 될 만한 상황들을 정리하여 외국인 고객들을 자신감 있게 대할 수 있도록 하는 내용으로 구성하였다.

부디 이 책을 통해 조금이나마 영어때문에 고민하는 학생들에게 도움이 될 수 있기를 소망하여 본다.

저자 올림

Contents

Theory of the hairdressing

Dialogue

Chapter 01

Jobs in Hairdressing & Beauty

Vocabularies

1. **Can't get enough of it** : 정말 좋아하다, 눈을 떼지 못하다
2. **Enthusiasm** : 열정, 열의
3. **Passion** : 열정, 정렬
4. **Variation** : 변화, 변형
5. **Career** : 직업
6. **Hairstylist** : 헤어스타일리스트
7. **Hairdresser** : 헤어드레서
8. **Interaction** : 상호작용, 관련
9. **Transition** : 변천, 추이, 변화
10. **Rapport** : 인간관계
11. **Anticipate** : (사람, 물음, 요구, 부탁 등에) 미리 응하다.
    시키기 전에 하다. 요구를 들어주다.
12. **Transformation** : 변신, 변화
13. **Repertory** : 레퍼토리
14. **Radical**: 극단적인, 근본적인
15. **Celebrity** : 유명인사, 명사
16. **Distribute** : 분산하다, 골고루 펴바르다.
17. **Evenly** : 골고루
18. **Damp** : 촉촉한, 살짝 젖은

Let's translate this!

♠ Being a hairdresser is a kind of career choice.

♠ Anyone can cut hair.

♠ But not anyone can be a good hairdresser.

♠ It is not chemical science to learn the techniques of cutting hair, but it takes so much more to be really good in this great profession.

♠ Let me begin however with the most important element.

♠ It is neither a skill nor a talent.

♠ It is the driving energy that makes us good at anything we do and without it, we would not even have fun working.

♠ It is passion.

♠ To be an excellent hairdresser you have to love what you do.

♠ If you dream it, live it and can't get enough of it, then you have the perfect start.

♠ Passion is the basis for any artistic and creative work.

♠ Without those inner sparks no great work can be come to life.

♠ And yes, you should be an artist.

♠ Even simple styles require a feeling for balance and a steady hand.

♠ Be a sculptor.

♠ Sculptors have to be able to see what they want to produce in a mass of unshaped material.

♠ As Michelangelo put it once: "I saw the angel in the marble and carved until I set him free."

♠ Do you see the style and finished look in the mass of hair?

♠ Be a color expert.

♠ Understand not only the basics but also the magic of mixing colors, combining color profiles, using light and shadow as well as interactions of warm and cold colors next to each other.

- For example when hair is colored in different tones and with contrasting hues.

- Be able to apply paint with a brush in a targeted and controlled way and to create smooth transitions between the shades.

- Be a designer.

- Some people are more creative than others who rather stay on the safe side and repeat what others have done before.

♠ Creativity is not only necessary to create the extravagant and post-modern looks.

♠ As a creative person you see things that others don't.

♠ It is as simple as that.

♠ In your mind you are able to visualize all kinds of shapes and colors and put them together.

♠ It is a big challenge to help someone to find the perfect look and to give advice.

♠ If you only have a small repertory of hairstyles that you are comfortable with and cannot think outside the box, your success will be limited as well.

♠ Be a great listener.

♠ No matter how creative or artistically and technically advanced you are, it won't do any good unless your customer is happy.

- And they are only happy when they leave the salon more beautiful than they came in and feel that all of their wishes were fulfilled.

- To understand the wants and needs of someone you have to listen carefully to what they say and also understand what they are not saying.

- This is where the next item on this list comes in:

- Be a psychologist.

♠ Know what to do with the information that you heard.

♠ Listen between the lines and understand the motivation of your customers.

♠ You have to be able to build a rapport with your customers and get to know them.

♠ Understand what they mean when they talk about the transformation of their hair, which often means a transformation of themselves.

♠ Especially women tend to visit hair salons with radical wishes like cutting long hair very short or applying an extreme color, when they actually want to change their life.

♠ Also, many people love to come to a salon just to get some personal attention and to talk to someone, be it about celebrity gossip or about more personal issues.

♠ This gives being a hairdresser a whole other dimension that is not mentioned in schools and during training, same as a bartender does not only pour drinks.

♠ A good one listens and anticipates.

♠ No matter what brings you to want to be a hairdresser, or what you expect from a hairdressing career, there are good and bad points about this path.

♠ Stylists are trained to look at more than just the hair when a client comes into the salon for services.

♠ This is because creating a flattering look requires more than just an evaluation of the hair.

♠ There are not many jobs that make it so easy to work freelance as a career in hair.

♠ Salons are perhaps the first thing that comes to mind but there are many other options and the limit is only your own imagination.

♠ The freedom you have being a mobile hairdresser is unbelievable.

♠ You don't feel tied down to anything, you work your own hours and your make your own money.

♠ Hairdressing insurance is cover that all hairdressers should have whether you are a self employed hairdresser or if you run a larger business with several employees.

♠ If you have beautiful hair you have a good chance to become a hair model. Most salons are looking for models so that they can practice the new looks.

Translation

♠ Being a hairdresser is a kind of career choice.

→ 헤어드레서가 된다는 것은 하나의 직업적 선택이다.

♠ Anyone could be able to cut hair.

→ 누구나 머리를 자를 수는 있을 것이다.

♠ But not anyone can be a good hairdresser.

→ 그러나 누구나 훌륭한 헤어드레서가 될 수는 없다.

♠ It is not extremely difficult science to learn the techniques of cutting and styling hair, but it takes so much more to be really good in this great profession.

→ 머리를 커트하고 스타일링 하는 것이 아주 어려운 과학같은 것은 아닐지라도 훌륭한 프로 스타일리스가 되려면 많은 것들이 요구되어진다.

♠ Let me begin however with the most important element.

→ 어쨌든 가장 중요한 요소부터 시작하여 보자.

♠ It is neither a skill nor a talent.

→ 그것은 기술도 아니고 재능도 아니다.

♠ It is the driving energy that makes us good at anything we do and without it, we would not even have fun working.

→ 우리가 무엇을 하던지 간에 잘 할 수 있도록 해주는 것은 바로 하고자하는 강한 에너지이다. 이것이 없으면 즐겁게 일할 수조차 없다.

♠ It is passion.

→ 그것은 열정이다.

♠ To be an excellent hairdresser you have to love what you do.

→ 정말 훌륭한 헤어드레서가 되려면 반드시 당신이 하는 일을 사랑해야만 한다.

♠ If you dream it, live it and can't get enough of it, then you have the perfect start

→ 당신이 만약 미용을 꿈꾸고, 함께 하며, 눈을 떼지 못할 만큼 좋아한다면 그거야 말로 완벽한 시작이라고 할 수 있다.

♠ Passion is the basis for any artistic and creative work.

→ 열정이야말로 예술적이고 창조적인 일들의 가장 중요한 기본이다.

♠ Without those inner sparks no great work can be come to life.

→ 내부의 강력한 스파크가 없이 인생을 활기차게 할 만한 일은 없다.

♠ And yes, you should be an artist.

➾ 그렇다. 당신은 아티스트가 되어야만 한다.

♠ Even simple styles require a feeling for balance and a steady hand.

➾ 아무리 단순한 스타일일지라도 균형과 꾸준한 손길이 요구된다.

♠ Be a sculptor.

➾ 조각가가 되어라.

♠ Sculptors have to be able to see what they want to produce in a mass of unshaped material.

➾ 조각가는 조각되지 않은 덩어리에서 어떠한 형상을 만들지를 볼 수 있어야 한다.

♠ As Michelangelo put it once: "I saw the angel in the marble and carved until I set him free."

➾ 미켈란젤로는 "나는 대리석 안에서 천사를 보았고 그것이 자유로워질 때까지 조각하였다"라고 하였다.

♠ Do you see the style and finished look in the mass of hair?

➾ 당신은 머리카락을 볼때 그 안에서 완성된 이미지와 스타일까지 볼 수 있는가?

♠ Be a color expert.

➾ 컬러 전문가가 되어라.

♠ Understand not only the basics but also the magic of mixing colors, combining color profiles, using light and shadow as well as interactions of warm and cold colors next to each other

⇨ 컬러에 대한 기본지식뿐만 아니라 컬러를 섞고, 색상들을 결합하며, 온색과 한색의 대비나 명도와 채도를 이용한 신비한 색상을 이해하여야 한다.

♠ For example when hair is colored in different tones and with contrasting hues.

⇨ 예를 들면 모발을 다른 색상이나 반대 색상으로 바꾸는 것이다.

♠ Be able to apply paint with a brush in a targeted and controlled way and to create smooth transitions between the shades.

⇨ 목표가 되는 부분에 붓을 이용해 부드럽게 도포해 주면서 다른 두색에서 부드럽게 변화되도록 한다.

♠ Be a designer.

⇨ 디자이너가 되어라.

♠ Some people are more creative than others who rather stay on the safe side and repeat what others have done before.

⇨ 그들은 안정된 쪽에만 머무르기 좋아하고 그전에 한것을 따라하는 것만 좋아하는 사람들보다는 더욱 창의적이다.

♠ Creativity is not only necessary to create the extravagant and post-modern looks.

⇨ 창조적이라고해서 반드시 사치스럽거나 포스트모던하고 전위적인 스타일을 만들어야 하는 것은 아니다.

♠ As a creative person you see things that others don't.

➾ 창조적인 사람은 남들이 보지 못한 것들을 볼 수 있다.

♠ It is as simple as that.

➾ 이것은 그처럼 단순한 것이다.

♠ In your mind you are able to visualize all kinds of shapes and colors and put them together.

➾ 당신의 마음속에서는 모든 종류의 스타일과 컬러들을 함께 합하여 상상하는 것이 가능하다.

♠ It is a big challenge to help someone to find the perfect look and to give advice.

➾ 어떤 사람에게 완벽한 스타일을 제안해주고 그러한 도움을 주는 것은 큰 도전이라고 할 수 있다.

♠ If you only have a small repertory of hairstyles that you are comfortable with and cannot think outside the box, your success will be limited as well.

➾ 만약 당신이 몇가지 헤어스타일만 할 수 있고 그것에 대해서 만족한다면 더 넓은 세상은 상상도 할 수 없을 것이며 당신의 성공 또한 제한될 것이다.

♠ Be a great listener.

➾ 좋은 청취자가 되어라.

♠ No matter how creative or artistically and technically advanced you are, it won't do any good unless your customer is happy.

⇨ 당신이 얼마나 창의적이거나 예술적이거나 기술적으로 앞서 있던지 간에 고객이 행복하지 않다면 그것은 절대 좋은 것이라고 할 수 없을 것이다.

♠ And they are only happy when they leave the salon more beautiful than they came in and feel that all of their wishes were fulfilled.

⇨ 또한 고객들이 살롱에 들어와서 그들이 바라는 것들을 이루고 살롱에 들어 올때보다 더 아름다운 모습으로 나갈때 그들은 행복하다.

♠ To understand the wants and needs of someone you have to listen carefully to what they say and also understand what they are not saying.

⇨ 고객의 바라는 욕구와 니드를 이해하기 위해서는 고객이 말하는 바를 주의깊게 들어야 하며 또한 그들이 말하지 않는 것 또한 이해하도록 하여야 한다.

♠ This is where the next item on this list comes in:

⇨ 이점이 바로 헤어드레서가 갖추어야 할 아래 덕목을 설명해주는 요점이다.

♠ Be a psychologist.

⇨ 심리학자가 되어라.

♠ Know what to do with the information that you heard.

⇨ 너가 들은 정보를 가지고 무엇을 할지를 알아야 한다.

♠ Listen between the lines and understand the motivation of your customers.

➩ 고객의 동기를 이해하고 말속에 숨어있는 의도를 들어야 한다.

♠ You have to be able to build a rapport with your customers and get to know them.

➩ 당신의 고객을 알고 고객과 라포를 형성할 수 있어야만 한다.

♠ Understand what they mean when they talk about the transformation of their hair, which often means a transformation of themselves.

➩ 고객이 그들 머리의 변화를 이야기할 때 사실은 종종 그들 자신의 변화를 원하는 경우도 많이 있다.

♠ Especially women tend to visit hair salons with radical wishes like cutting long hair very short or applying an extreme color, when they actually want to change their life.

➩ 특별히 여성고객의 경우는 긴머리를 짧게 자르거나 극단적인 컬러를 하려고 할 때에는 그들의 삶자체를 바꾸고 싶은 경우가 많다.

♠ Also, many people love to come to a salon just to get some personal attention and to talk to someone, be it about celebrity gossip or about more personal issues.

➩ 또한 많은 사람들이 유명인사들의 가십거리나 사적인 이야기를 나누기 위해서 또는 다른 사람들의 주목을 받고 싶어서 헤어살롱에 들리는 것을 좋아한다.

♠ This gives being a hairdresser a whole other dimension that is not mentioned in schools and during training, same as a bartender does not only pour drinks.

⇒ 이런 것은 학교나 실습과정에서 배우는 것이 아니기에 여러가지 측면에서 헤어드레서가 하는 일을 생각해 보아야 하며 이것은 마치 바텐더가 단지 음료를 따르는 일만 하는 직업이 아닌 것과 비슷하다.

♠ A good one listens and anticipates.

⇒ 훌륭한 헤어드레서는 잘 듣고 고객을 만족시킬 수 있다.

♠ No matter what brings you to want to be a hairdresser, or what you expect from a hairdressing career, there are good and bad points about this path.

⇒ 여러분이 왜 헤어드레서가 되려고 했는지 헤어드레서가 되어 무엇을 기대 하던지간에 미용이라는 직업은 좋은 점과 나쁜 점이 있다.

♠ Stylists are trained to look at more than just the hair when a client comes into the salon for services.

⇒ 스타일리스트는 고객이 샵에 들어올 때 헤어뿐만 아니라 고객의 전체적인 스타일까지 살펴보아야한다.

♠ This is because creating a flattering look requires more than just an evaluation of the hair.

⇒ 고객에게 어울리는 스타일을 제안해 주려면 단지 헤어스타일만 보아서는 안된다.

♠ There are not many jobs that make it so easy to work freelance as a career in hair.

⇒ 헤어드레서만큼 프리랜서로 일할 수 있는 일도 많지 않다.

♠ Salons are perhaps the first thing that comes to mind but there are many other options and the limit is only your own imagination.

⇨ 헤어살롱이 가장 먼저 떠오르겠지만 샵말고도 여러 다른 일들이 있으며, 그 종류도 무궁무진하다.

♠ The freedom you have being a mobile hairdresser is unbelievable.

⇨ 프리랜서로 일하는 헤어디자이너의 자유는 정말 놀랄만큼 좋은 일이다.

♠ You don't feel tied down to anything, you work your own hours and your make your own money.

⇨ 어디에 얽매이는것 없이 당신의 시간에 일한만큼 돈을 벌 수 있다.

♠ Hairdressing insurance is cover that all hairdressers should have whether you are a self employed hairdresser or if you run a larger business with several employees.

⇨ 미용사면허는 혼자서 일을 하든 사람을 고용하여 샵을 경영하든 반드시 필요하다.

♠ If you have beautiful hair you have a good chance to become a hair model. Most salons are looking for models so that they can practice the new looks.

⇨ 만약 여러분 중에 아름다운 모발을 가지고 있다면 헤어모델이 되는 기회를 가져보는 것도 좋다. 대부분의 살롱에서는 새로운 스타일을 시도해보기 위해 모델을 찾고 있다.

Chapter 02

Terminology for Shampoo, Rinse, Conditioning

Vocabularies

1. **Shampoo** : 샴푸하다.
2. **Rinse** : 물로 헹구다. 씻다.
3. **Hair Conditioner** : 헤어컨디셔너. 국내에서는 린스라고 하는 것의 영어식 바른표현
4. **Wash** : 씻다.
5. **Sebum** : 피지
6. **Dandruff** : 비듬
7. **Towel dry** : 타월 드라이. 모발세정 후 타월로 물기를 가볍게 제거하는 것을 말한다.
8. **Basin** : 세면기. 세면대
9. **Texture** : 질감. 결, 헤어드레싱에서는 모발이 얼마나 곱슬거리는지, 웨이브가 많은지, 스트레이트한지를 나타낸다고 생각하면 이해가 빠를 것이다.
10. **Manageability** : 관리
11. **Detergent** : 세제
12. **component** : 성분. 구성요소
13. **Replacement** : 보충, 대체
14. **Remove** : 제거하다.
15. **apply** : 도포하다. 바르다.
16. **improve** : 개선시키다. 향상시키다. 더욱 좋게 하다.
17. **surfactant** : 계면활성제
18. **lather** : 거품을 내다.
19. **Tension** : 텐션
20. **Nape** : 네이프
21. **Dry shampoo** : 드라이 샴푸

Let's translate this!

♠ Hair washing is the cosmetic act of keeping hair clean by washing it with shampoo and water.

♠ Hair conditioner may also be used to improve hair's texture and manageability.

♠ Two-in-one shampoos, which have both detergent and conditioning components, are now commonly also used as a replacement for shampoo and conditioner.

♠ To remove the sebum from the hair, some people apply a surfactant, most usually shampoo, but sometimes soap to their hair and lather the surfactant with water.

♠ he surfactant is rinsed out with water along with the dirt and dust.

♠ There are also dry shampoos, which are powders which remove sebum from the hair by soaking it up prior to being combed out.

Translation

♠ Hair washing is the cosmetic act of keeping hair clean by washing it with shampoo and water.

⇒ 헤어와싱은 샴푸와 물로 깨끗이 씻어내는 미용의 가장 기초적인 시술이다.

♠ Hair conditioner may also be used to improve hair's texture and manageability.

⇒ 헤어컨디셔너는 모발의 결을 더 좋게 해주고 관리하기 좋도록 해준다.

♠ Two-in-one shampoos, which have both detergent and conditioning components, are now commonly also used as a replacement for shampoo and conditioner.

⇒ 샴푸린스겸용제품은 세제 및 컨디셔닝 할 수 있는 성분이 들어있는 제품이다.

♠ To remove the sebum from the hair, some people apply a surfactant, most usually shampoo, but sometimes soap to their hair and lather the surfactant with water.

⇒ 모발의 피지를 제거하기 위해서 대부분 샴푸를 사용하지만 비누를 사용하거나 다른 계면활성제를 이용하여 거품을 내기도 한다.

♠ The surfactant is rinsed out with water along with the dirt and dust.

⇒ 계면활성제는 더러움과 먼지와 함께 물로 헹구어서 씻겨진다.

♠ There are also dry shampoos, which are powders which remove sebum from the hair by soaking it up prior to being combed out.

⇒ 또한 드라이 샴푸는 파우더로 피지를 제거하는 것으로서 빗으로 빗기 전에 파우더를 충분히 도포해주어야 한다.

드라이샴푸

tip 계면활성제와 세제의 차이

계면이란 성질이 다른 두 가지 물질이 서로 접하고 있을 때 그 경계면을 말한다. 이러한 계면은 액체와 액체, 액체와 기체, 액체와 고체, 고체와 고체간에도 존재한다. 계면활성제는 독특한 화학구조를 가지고 있다. 분자의 한쪽에 물과 친화력이 큰 친수기가 있고 다른 한쪽에는 기름과 친화성이 큰 친유기를 가지고 있다. 대표적 계면활성제가 비누이다.

계면에 흡착하여 그 표면장력을 감소시키는 물질로서 표면활성제라고도 한다. 비누는 그 대표적인 것으로, 비눗물의 표면장력은 물에 비하여 훨씬 작다.

Chapter 03

Terminology for Haircutting

Vocabularies

1. **Guide** : 가이드
2. **Blocking** : 브로킹
3. **Section** : 섹션
4. **Crown** : 크라운.
5. **Forehead** : 이마
6. **Flattering** : 실물보다 더욱 돋보이게 하는.
7. **Facial** : 얼굴의
8. **Executed** : 실시한, 시행한.
9. **Slant** : 비스듬히 경사를 주다. 굴곡.
10. **Comb** : 빗
11. **Horizontal** : 가로
12. **Vertical** : 세로
13. **Diagonal** : 사선, 대각선의
14. **Pivot** : 방사선의
15. **Blunt cut** : 모발을 직선으로 자르는 것을 말한다. 무게감을 주는 라인을 만들기 위해 모든 모발이 한 곳에 위치하도록 자르게 된다.
16. **Bob** : 단발
17. **Wet cut** : 모발을 샴푸잉하거나 스프레이를 분무하여 젖은 상태로 커트하는 방법
18. **Dry cut** : 마른 모발에서 커트하는 방법.
19. **Razor** : 레이저
20. **Clipper** : 클리퍼

21. **Fringe** : 앞머리

22. **Bang** : 뱅, 앞머리

23. **Shape** : 모양, 쉐이프

24. **one lengh cut** : 원랭스 커트

25. **Graduation cut** : 그라데이션 커트

26. **Layer cut** : 레이어 커트

27. **Towel dry** : 타월 드라이

28. **Tangle** : 엉킴

29. **Extend** : 연장하다.

30. **Cresent** : 반달

31. **Forehead** : 이마

32. **Distribute** : 분배하다. 분산하다.

33. **Evenly** : 골고루

34. **Damp** : 적신.

35. **flat-iron** : 플랫 아이롱. 매직기

36. **Edge** : 가장자리

37. **Procedure** : 시술, 절차

Let's translate this!

♠ The Blunt Bob haircut is one of the most classic looks in history.

♠ And It is flattering to many facial types in one length or another.

♠ It can be worn with or without shorter fringebangs), and many people who wish to shorten a long forehead prefer to have fringe.

♠ Some people who like the blunt style of haircut don't like the hard line created by the basic blunt style.

- One easy way to get a softer, smoother turn on the style is the use of a razor tool to texture the ends of the cut.

- We've adapted our previously executed blunt bob style here with bangs and razor.

- Cutting procedure of the bob style with bang and razor.

- Perform the Basic Blunt Bob Haircut, and re-wet the hair in preparation for texturing.

♠ If you are working with a previously executed bob cut, shampoo and condition the hair, towel dry, and comb the hair to remove any tangles.

♠ Begin by sectioning off the fringe area of the hair.

♠ This section can take any shape you prefer, but it generally extends no wider than the outer corners of the eyes.

♠ Making the fringe area a crescent shape above the forehead, approximately 1 inch deep at its widest pointdirectly above the nose).

♠ The fringe was then cut straight across following a horizontal line.

♠ To begin the texturing, we start with the bangs.

♠ Comb the bangs together from both the right and left sides, to assure they are centered.

♠ Then, use the razor angled slightly against the hair and gently 'shave' the last third of the hair.

♠ The goal is to create gentle slant on the ends of the bangs, to soften the line. Remember that you can always go back and cut more hair, but once it's gone, it's gone.

♠ We then move to the blunt line of the haircut.

- Comb out segments of hair approximately the width of the razor, and hold them with light tension between the index and middle fingers of one hand.

- Use the razor as directed above and 'shave' the bottom inch of the cut line.

- Work your way around the head in this manner until you have razor-cut all along the bottom edge of the cut.

♠ We styled our demonstration using a small amount of gel evenly distributed throughout the still-damp hair.

♠ Using a blow-dryer on a high-heat, low-speed setting and a round ceramic brush, the hair was dried and the ends curled gently under.

♠ A light misting of hairspray and a flat-iron tool was used to reinforce the straight style.

- Afterwards, a small amount of pomade with shine enhancer was used to smooth the finished style and define the part in the style.

- The finished result is a style with softer lines than the traditional blunt bob hair-style.

- The addition of bangs makes it an ideal look for anyone who needs to shorten a long face, or mask a high forehead.

Translation

♠ The Blunt Bob haircut is one of the most classic looks in history.

⇨ 직선적인 단발커트는 역사적으로 가장 클래식한 스타일 중의 하나이다.

♠ And It is flattering to many facial types in one length or another.

⇨ 원랭스 또는 다른 형태의 단발 스타일은 많은 얼굴형을 돋보이게 하여준다.

♠ It can be worn with or without shorter fringebangs), and many people who wish to shorten a long forehead prefer to have fringe.

⇨ 단발머리는 앞머리와 함께 종종 커트되어지고, 이마가 긴 얼굴형에는 종종 짧은 앞머리를 원하는 사람이 많다.

♠ Some people who like the blunt style of haircut don't like the hard line created by the basic blunt style.

⇨ 블런트 커트를 원하는 경우에 너무 뭉뚝한 라인을 원하는 사람은 그다지 많지 않다.

♠ One easy way to get a softer, smoother turn on the style is the use of a razor tool to texture the ends of the cut.

⇨ 가볍고 부드러운 느낌을 얻는 가장 쉬운 방법은 커트의 끝부분을 레이저를 이용하여 질감처리를 해주는 것이다.

♠ We've adapted our previously executed blunt bob style here with bangs and razor.

⇨ 여기에 레이저와 앞머리를 이용하여 시술한 커트를 소개하고자 한다.

♠ Cutting procedure of the bob style with bang and razor

⇨ 레이저를 이용한 앞머리가 있는 단발커트 시술 순서

♠ Perform the Basic Blunt Bob Haircut, and re-wet the hair in preparation for texturing.

⇨ 기본적인 블런트 단발 커트를 시술하기 위해서 질감처리를 위한 준비로서 물을 분무한다.

♠ If you are working with a previously executed bob cut, shampoo and condition the hair, towel dry, and comb the hair to remove any tangles.

⇨ 만약 이미 단발 커트를 한 머리라면 샴푸와 컨디셔닝을 한 뒤 타월드라이하고 빗질을 잘하여 머리에 엉킴이 없도록 준비한다.

♠ Begin by sectioning off the fringe area of the hair.

⇨ 모발의 앞머리 부분은 섹셔닝하여 둔다.

♠ This section can take any shape you prefer, but it generally extends no wider than the outer corners of the eyes.

➾ 섹션을 나눌 때는 원하는 모양으로 나눌 수 있지만 대개는 양쪽 눈가의 가장 자리보다 더 넓게 나누지는 않는다.

♠ Making the fringe area a crescent shape above the forehead, approximately 1 inch deep at its widest pointdirectly above the nose).

➾ 이마 위쪽으로 반달모양의 섹션을 나눈다. 코의 위쪽방향으로 약 2.5㎝ 똑바로 올라간 부위가 가장 넓은 부위가 되도록 나눈다.

♠ The fringe was then cut straight across following a horizontal line.

➾ 가로 방향으로 직선으로 커트해 나간다.

♠ To begin the texturing, we start with the bangs.

➾ 앞머리부터 질감처리를 해준다.

♠ Comb the bangs together from both the right and left sides, to assure they are centered.

➾ 오른쪽에서 부터 왼쪽을 빗질하여 정확하게 균형이 맞도록 주의한다.

♠ Then, use the razor angled slightly against the hair and gently 'shave' the last third of the hair

➾ 그리고 나서 약간의 각도를 주어 모발의 아래 3분의 1지점을 레이저로 질감 처리한다.

The goal is to create gentle slant on the ends of the bangs, to soften the line. Remember that you can always go back and cut more hair, but once it's gone, it's gone.

➾ 이렇게 함으로써 앞머리에 부드러운 느낌을 만들 수 있다. 모발은 한번 잘려나가면 회복이 불가능하기 때문에 조금씩 작업해 나가야 한다.

We then move to the blunt line of the haircut.

➾ 앞머리를 마치고 나면 다른 부분도 진행하도록 한다.

Comb out segments of hair approximately the width of the razor, and hold them with light tension between the index and middle fingers of one hand.

➾ 모발을 레이저의 넓이만큼 적당한 양으로 나눈 뒤에 검지와 중지 손가락으로 가볍게 잡아준다.

Use the razor as directed above and 'shave' the bottom inch of the cut line.

➾ 커트선의 아래쪽의 약 2.5cm 부분을 레이저를 이용하여 스트록한다.

Work your way around the head in this manner until you have razor-cut all along the bottom edge of the cut.

➾ 커트선의 가장자리를 위의 방식으로 모두 진행해 나간다.

We styled our demonstration using a small amount of gel evenly distributed throughout the still-damp hair.

➾ 아직 조금 물기가 남은 모발에 적은양의 젤을 골고루 도포한 뒤 스타일링 한다.

♠ Using a blow-dryer on a high-heat, low-speed setting and a round ceramic brush, the hair was dried and the ends curled gently under.

⇒ 고온의 헤어드라이어로 천천히 바람을 주면서 세라믹 브러쉬로 원을 그리듯이 빗어가면서 모발의 끝이 자연스럽게 아래방향으로 커브를 그리듯이 말려준다.

♠ A light misting of hairspray and a flat-iron tool was used to reinforce the straight style.

⇒ 가벼운 질감의 미스트를 분사한 뒤 매직기를 이용하여 스트레이트한 스타일을 더욱 강조하여 준다.

♠ Afterwards, a small amount of pomade with shine enhancer was used to smooth the finished style and define the part in the style.

⇒ 그런 뒤에 소량의 포마드를 도포하여 가르마를 잡아주고 매끈하게 마무리한다.

♠ The finished result is a style with softer lines than the traditional blunt bob hair-style.

⇒ 피니싱 스타일은 전통적인 블런트 단발 커트에 비하여 더욱 부드럽다.

♠ The addition of bangs makes it an ideal look for anyone who needs to shorten a long face, or mask a high forehead.

⇒ 앞머리는 얼굴이 긴형이거나 긴 이마를 보완하고 싶은 사람에게 잘 어울린다.

Chapter 04

Terminology for Braiding Hair

Vocabularies

1. **Scalp** : 두피
2. **Wearer** : 착용자
3. **Braid** : 땋은머리
4. **Plait** : 땋은 머리
5. **circulate** : 순환되다.
6. **invisible** : 보이지않는
7. **Thin** : 얇은
8. **Slice** : 슬라이스
9. **Combine** : 결합시키다. 합치다.
10. **Strand** : 가닥
11. **Shift** : 변화, 전환 ,이동, 옮기다, 바뀌다.

Let's translate this!

♠ A popular style for summer for women with long hair because it allows the air to circulate to the scalp and helps to keep the wearer cool.

♠ French Braid

♠ Styling techniques for a French Braid, also called Invisible Braid or French Plait.

♠ Step One:

Start with clean hair.

♠ The hair can be freshly washed, conditioned and towel-dried, or can be fully dry, as you prefer.

♠ I prefer to braid damp hair.

♠ Use a detangling spray or leave-in conditioner and comb through the hair until all tangles are removed.

♠ Comb the hair back and away from the face.

♠ Section a triangular wedge of hair, starting at the forehead and extending back toward the crown of the head.

♠ This will be the starting point of your braid.

♠ The model have curly hair, which makes braiding easier in many ways.

♠ If you have trouble braiding the hair you are working on, you can try adding some styling productmousse, gel, setting lotion, etc. to add some texture and "grip" to the hair.

♠ Step Two :

Divide the triangular section into three 'strands' and hold them in one hand, separated by the fingers of the holding hand.

♠ Begin braiding by crossing the right strand over the center strand, carefully shift the strands to the opposite hand, and then cross the left strand over the 'new' center strand.

♠ Shift the strands back to the original hand.

♠ Step Three :

Continue the braid by taking a thin slice of hair from the right side of the head and combine it with the right strand, then cross this over the center strand.

- Shift the strands to the opposite hand.

- Now, take a slice from the left side of the head and combine it with the left strand and cross it over the center strand.

- Return the strands to the starting hand.

- Step Four :

  Repeat step three until you have made your way down the head to the nape of the neck.

♠ By maintaining the tension with which you hold the hair, you can make the braid as tight or as loose as you want.

♠ Step Five :

Once you've reached the nape of the neck, you can finish the braid however you want.

♠ You can gather the hair in an elastic band at the neck and leave the remaining hair to flow freely, or you can continue crossing the strands right-over-center and left-over-center until you reach the ends of the hair.

- If you started with damp hair, or find the hair is damp from the addition of detangler or styling product and you prefer the dry look, you can simply use a bonnet dryer or a blow-dryer with a diffuser attachment to dry the braided hair.

- Once dry, the "product" used can be softened by carefully pressing the hair with your fingers to separate the hair.

- For the demonstration, I continued the braid for the full length of the hair.

♠ The finished braid can be decorated with clips, ribbons, bows, flowers, etc.

♠ Virtually anything could be used to decorate or add additional interest to the braid.

♠ The Invisible Braid finishes as a smooth, tidy look, holds well, and is an excellent way to wear long hair when you are engaging in all kinds of active endeavors, like sporting events, swimming, housework, etc.

프렌치 브레이드

Translation

♠ A popular style for summer for women with long hair because it allows the air to circulate to the scalp and helps to keep the wearer cool.

⇨ 머리를 땋는 것은 여름철 긴 머리를 가진 여성들에게 매우 인기 있는 스타일인데 그 이유는 바람이 두피 안으로 잘 통하게 하여주기 때문이다.

♠ French Braid

⇨ 프렌치 브레이드

♠ Styling techniques for a French Braid, also called Invisible Braid or French Plait.

⇨ 프렌치 브레이드는 인비져블 브레이드 또는 프렌치 플레이트라고 한다.

♠ Step One: Start with clean hair.

⇨ 먼저 모발을 깨끗하게 한다.

♠ The hair can be freshly washed, conditioned and towel-dried, or can be fully dry, as you prefer.

⇨ 모발은 깨끗하게 샴푸한 뒤에 타월드라이를 하거나 완전히 말린다.

♠ I prefer to braid damp hair.

⇨ 나는 개인적으로 약간 젖은 모발을 선호한다.

♠ Use a detangling spray or leave-in conditioner and comb through the hair until all tangles are removed.

⇨ 뿌리는 타입의 컨디셔너나 엉킴방지 스프레이를 뿌린 뒤 빗으로 빗어 엉킴을 제거한다.

♠ Comb the hair back and away from the face.

⇨ 얼굴로부터 뒷방향으로 빗질한다.

♠ Section a triangular wedge of hair, starting at the forehead and extending back toward the crown of the head.

⇨ 이마 뒤쪽으로 크라운쪽으로 삼각형의 섹션을 뜬다.

♠ This will be the starting point of your braid.

⇨ 여기에서부터 브레이딩이 시작하게 된다.

♠ The model have curly hair, which makes braiding easier in many ways.

⇨ 모델이 곱슬머리일 경우 여러모로 브레이딩 시술이 수월하다.

♠ If you have trouble braiding the hair you are working on, you can try adding some styling productmousse, gel, setting lotion, etc. to add some texture and "grip" to the hair.

➾ 만약에 작업하기가 조금 어렵다면 무스, 젤, 세팅로션 등의 스타일링 제품을 사용하여 텍스츄어를 만들어서 시술하여도 된다.

♠ Step Two :

Divide the triangular section into three 'strands' and hold them in one hand, separated by the fingers of the holding hand.

➾ 삼각형의 섹션으로 나눈 모발 스트렌드를 세가닥으로 나누어서 한손으로 잡고, 잡은 손의 손가락으로 세가닥을 나누어준다.

♠ Begin braiding by crossing the right strand over the center strand, carefully shift the strands to the opposite hand, and then cross the left strand over the 'new' center strand.

➾ 오른쪽 가닥을 중간가닥으로 교차한 뒤 조심스럽게 반대쪽 손으로 바꾸어주고 왼쪽가닥을 새롭게 생긴 중간 가닥위로 교차한다.

♠ Shift the strands back to the original hand.

➾ 뒤의 모발을 원래 손쪽으로 옮긴다.

♠ Step Three :

Continue the braid by taking a thin slice of hair from the right side of the head and combine it with the right strand, then cross this over the center strand.

➾ 오른쪽에서 약간의 슬라이스를 떠서 잡은 뒤 오른쪽 가닥과 합치고 이것을 중앙의 가닥위로 겹치면서 땋아나간다.

♠ Shift the strands to the opposite hand.

⇒ 모발을 반대쪽 손으로 바꿔 잡는다.

♠ Now, take a slice from the left side of the head and combine it with the left strand and cross it over the center strand.

⇒ 이제 두상의 왼쪽에서 슬라이스를 떠서 왼쪽 가닥과 합친 뒤에 중앙의 가닥위로 겹친다.

♠ Return the strands to the starting hand.

⇒ 머리가닥을 처음 시작한 손으로 돌아간다.

♠ Step Four :

Repeat step three until you have made your way down the head to the nape of the neck.

⇒ 머리에서부터 모발 끝까지 이러한 과정을 반복한다.

♠ By maintaining the tension with which you hold the hair, you can make the braid as tight or as loose as you want.

⇒ 모발을 잡을 때 균일한 텐션을 유지해야 느슨한 형태 또는 단단한 형태의 브레이드를 얻을 수 있다.

♠ Step Five :

Once you've reached the nape of the neck, you can finish the braid however you want.

⇒ 목덜미정도까지 다다르면 브레이드를 마무리할 수 있다.

♠ You can gather the hair in an elastic band at the neck and leave the remaining hair to flow freely, or you can continue crossing the strands right-over-center and left-over-center until you reach the ends of the hair.

⇨ 모발은 목주변에서 고무줄로 묶어서 남은 머리를 자유롭게 내려두거나 오른쪽과 왼쪽의 모발을 중앙쪽으로 교대해 가면서 모발끝까지 땋아도된다.

♠ If you started with damp hair, or find the hair is damp from the addition of detangler or styling product and you prefer the dry look, you can simply use a bonnet dryer or a blow-dryer with a diffuser attachment to dry the braided hair.

⇨ 젖은 머리나 또는 스타일링 제품을 이용하여 수분을 줄 수도있고, 만약 건조한 느낌으로 연출하고 싶다면 여러 종류의 드라이어를 이용하여 땋은 머리를 말릴 수도 있다.

♠ Once dry, the "product" used can be softened by carefully pressing the hair with your fingers to separate the hair.

⇨ 모발이 건조되면 제품을 이용하여 조심스럽게 어루만져 모발을 잘 정리할 수 있다.

♠ We continue the braid for the full length of the hair.

⇨ 모발 전체를 땋으며 계속 진행한다.

♠ The finished braid can be decorated with clips, ribbons, bows, flowers, etc.

⇨ 다 땋은 머리에는 클립이나 리본 꽃등을 이용하여 장식할 수 있다.

♠ Virtually anything could be used to decorate or add additional interest to the braid.

⇒ 사실상 브레이드의 장식을 위해서는 모든 재료가 사용될 수 있다.

♠ The Invisible Braid finishes as a smooth, tidy look, holds well, and is an excellent way to wear long hair when you are engaging in all kinds of active endeavors, like sporting events, swimming, housework, etc.

⇒ 프렌치 브레이드는 부드럽고 단정하며 단단하게 고정되어 긴머리를 가진 분에게 운동, 수영, 여러가지 가사일 등의 활동적인 일들을 할 때도 적당하다.

Chapter 04

Terminology for wig

Vocabularies

1. **artificial** : 인공의 ,인조의
2. **covering** : 덮개, 막
3. **interwoven** : 서로 엮어서 짠, 짜넣은, 섞어서 짠
4. **synthetic** : 인조의
5. **full-cap** : 풀캡
6. **partial** : 부분적인, 부분의
7. **hand-tied** : 수작업한
8. **machine-tied** : 기계로 작업한
9. **factor** : 요인, 원인, 요소
10. **affect** : 영향을 주다
11. **fuss** : 소란, 불편함
12. **wig** : 가발
13. **wiglet** : 부분 가발

Let's translate this!

♠ A wig is an artificial covering for the head consisting of an interwoven network of hair.

♠ There are men's and women's styles, and the hair can be human or synthetic.

♠ The style can be full-cap, open-cap or partial coveragewiglet).

♠ A wig can be hand-tied or machine-tied. All of these factorsas well as others) can affect the price of the wig. Here is some basic information on different wig types:

- Partial Coverage Wigs: Also known as wiglets and hair pieces, partial coverage wigs are designed to add hair to specific areas of the head, or to add to a specific hairstyle.

- These hairpieces can either be built on mesh foundationslike many full wigs) or simply consist of wefts of hair that have been attached to combs or elastics for placement onto the existing hair.

- They are great for creating hairstyles that give the illusion of longer hair without the added fuss of a full-sized wig.

♠ Full-Cap Wigs:

♠ Cap wigs are made using an elasticized mesh base and are usually hand-tied.

♠ They are typically denser and are great for people who have little or no hair of their own.

♠ The front edge of cap wigs tend to be made with material that resembles the client's scalp and will be under-knotted to hide the base of the wig, allowing for natural-looking results.

♠ Open-Cap Wigs:

♠ Also called cap-less wigs, open cap wigs are usually machine made wigs consisting of wefts of hair that have been stitched in a circular pattern onto elastic strips to form the cap-like shape for wear on the head.

♠ The biggest difference in full-cap and open-cap wigs is the existence of open spaces in the open-capcap-less) design.

♠ This makes the wig a good choice for women who want a wig that will blend with their natural hair, as the natural hair can be pulled through the spaces for a fuller, thicker-looking head of hair.

♠ Open-cap wigs are not recommended for women with very thin or no hair as the scalp will be visible between the wefts of the wig.

♠ Men's Wigs:

♠ Even men often opt to wear wigs and hairpieces to cover balding areas.

♠ Partial coverage wigshairpieces) for men are called toupees.

♠ These are the most common of men's hairpieces, although there are specially-designed full-cap wigs in men's styles for those men who have complete hair loss or prefer total coverage to the partial coverage of the toupee design.

♠ Hand-Tied Wigs:

♠ In hand-tied wigs and hairpieces, the hairs are individually knotted to the mesh cap by hand offering a more-natural look.

- Hand-tying also has the benefit of generally being more-loosely attached and therefore can be shifted along the mesh to alter the partings and styles more easily.

- This process takes longer to create the wig and therefore increases the price.

- Machine-Tied Wigs:

- Machine-tied wigs use wefts of hair that are sewn to a cap to create the wig or hairpiece more quickly.

♠ These wigs tend to be less expensive and more widely available.

♠ They are excellent for persons who need to be cost conscious in choosing their wigs and hairpieces.

♠ Semi-Hand-tied Wigs:

♠ These wigs feature a combination of machine sewn wefts and hand-tied hairs, as well as a combination of natural and synthetic hair, offering a durable, more-natural looking wig at a reasonable price.

Translation

♠ A wig is an artificial covering for the head consisting of an interwoven network of hair.

⇒ 가발은 모발을 엮어서 짠 망으로, 모발에 쓰기위해 만든 인조덮개이다.

♠ There are men's and women's styles, and the hair can be human or synthetic.

⇒ 남성용, 여성용, 인모, 인조모로 된 제품들이 있다.

♠ The style can be full-cap, open-cap or partial coverage(wiglet).

⇒ 풀캡, 오픈캡, 부분 가발 등의 종류가 있다.

♠ A wig can be hand-tied or machine-tied.

⇒ 수제작업 또는 기계 작업으로 만들어진다.

♠ Factiors affect the price of the wig.

→ 이러한 모든 요소들이 가발의 가격에 영향을 준다.

♠ Here is some basic information on different wig types:

→ 아래에 위그 종류에 따른 기본 정보를 정리해보았다.

♠ Partial Coverage Wigs:

→ 부분가발

♠ Also known as wiglets and hairpieces, partial coverage wigs are designed to add hair to specific areas of the head, or to add to a specific hairstyle.

→ 위글렛 또는 헤어피스라고도 불리어지는 부분가발은 특정 부분에 모량을 더해주거나 특별한 헤어스타일을 만들기 위해서 사용되어진다.

♠ These hairpieces can either be built on mesh foundationslike many full wigs) or simply consist of wefts of hair that have been attached to combs or elastics for placement onto the existing hair.

→ 이러한 헤어피스는 매쉬망 위에 심어지거나 기존의 모발에 고무줄이나 빗살의 형태를 달아서 고정하는 형태가 있다.

♠ They are great for creating hairstyles that give the illusion of longer hair without the added fuss of a full-sized wig.

→ 이러한 형태의 가발들은 전체가발을 사용하는 번거로움 없이 조금 더 긴 모발로 만들어줄 수 있다.

♠ Full-Cap Wigs:

⇒ 전체가발

♠ Cap wigs are made using an elasticized mesh base and are usually hand-tied.

⇒ 캡가발은 대게 신축성이 있는 메쉬 베이스에 수제공법으로 만들어진다.

♠ They are typically denser and are great for people who have little or no hair of their own.

⇒ 특히 이러한 가발은 조금 더 촘촘하여 모발 숱이 없거나 적은 사람에게 좋다.

♠ The front edge of cap wigs tend to be made with material that resembles the client's scalp and will be under-knotted to hide the base of the wig, allowing for natural-looking results.

⇒ 캡의 앞부분은 고객의 두피와 비슷한 것으로 하기 위해서 베이스 아래로 너팅을 하여 최대한 자연스럽게 보이도록 한다.

♠ Open-Cap Wigs:

⇒ 오픈 캡 위그

♠ Also called cap-less wigs, open cap wigs are usually machine made wigs consisting of wefts of hair that have been stitched in a circular pattern onto elastic strips to form the cap-like shape for wear on the head.

⇒ 또한 캡리스 위그라고도 불리는 오픈 캡 위그는 두상위에 쓰는 모자같은 형태에 고무밴드로 둥굴게 연결한 뒤 모발 웨프트를 달아서 만들었다.

♠ The biggest difference in full-cap and open-cap wigs is the existence of open spaces in the open-capcap-less) design.

⇨ 풀캡과 오픈캡의 가장 큰 차이점은 오픈되어 있는 공간의 유무에 있다.

♠ This makes the wig a good choice for women who want a wig that will blend with their natural hair, as the natural hair can be pulled through the spaces for a fuller, thicker-looking head of hair.

⇨ 가발의 틈으로 본래의 모발을 밖으로 빼내어 더욱 풍성하고 자연스럽게 연출 할 수 있다.

♠ Open-cap wigs are not recommended for women with very thin or no hair as the scalp will be visible between the wefts of the wig.

⇨ 오픈 캡은 모발이 매우 가늘거나 두피에 모발이 거의 없는 여성분에게는 그다지 추천할 만하지 않다.

♠ Men's Wigs:

⇨ 남성용 가발

♠ Even men often opt to wear wigs and hairpieces to cover balding areas.

⇨ 또한 남자들도 탈모부위를 가리기 위해서 헤어피스를 사용하는 경우도 있다.

♠ Partial coverage wigshairpieces) for men are called toupees.

⇨ 남자들을 위한 부분 가발을 투페라고 부른다.

♠ These are the most common of men's hairpieces, although there are specially-designed full-cap wigs in men's styles for those men who have complete hair loss or prefer total coverage to the partial coverage of the toupee design.

➾ 전두탈모나 부분탈모를 가진 남성분들을 위한 풀캡 가발도 있지만, 투페는 남자분들이 사용하는 헤어피스중에 가장 대중적인 스타일이다.

♠ Hand-Tied Wigs:

➾ 수제가발

♠ In hand-tied wigs and hairpieces, the hairs are individually knotted to the mesh cap by hand offering a more-natural look.

➾ 수제가발과 헤어피스는 더욱 자연스러운 모습을 얻기 위해서 매쉬캡에 한가닥씩 수제를 하고 있다.

♠ Hand-tying also has the benefit of generally being more-loosely attached and therefore can be shifted along the mesh to alter the partings and styles more easily.

➾ 수제 너팅은 대게 좀더 느슨하여, 가마나 스타일을 좀더 쉽게 바꿀수 있게 메쉬를 따라 변화를 줄 수 있다.

♠ This process takes longer to create the wig and therefore increases the price.

➾ 이러한 공정은 가발을 만드는데 시간이 더 걸리므로 가격이 더 높다.

♠ Machine-Tied Wigs:

➾ 기계수제가발

♠ Machine-tied wigs use wefts of hair that are sewn to a cap to create the wig or hairpiece more quickly.

⇒ 기계수제가발은 모발의 묶음으로써 더욱 빨리 가발을 만들어 내기 위해서 캡에 박음질로 연결한다.

♠ These wigs tend to be less expensive and more widely available.

⇒ 이런 종류의 가발은 더 저렴하고 널리 사용되어진다.

♠ They are excellent for persons who need to be cost conscious in choosing their wigs and hairpieces.

⇒ 이러한 제품은 가발을 선택할 때 가격에 관심이 있는 고객들에게 좋은 제품이다.

♠ Semi-Hand-tied Wigs:

⇒ 반수제 가발

♠ These wigs feature a combination of machine sewn wefts and hand-tied hairs, as well as a combination of natural and synthetic hair, offering a durable, more-natural looking wig at a reasonable price.

⇒ 이러한 형태는 기계와 수제가발의 결합형태로써, 합리적인 가격에 내구성도 오래가도록 된 형태이다.

♠ Synthetic Hair Wigs:

⇒ 인조가발

♠ Synthetic hair wigs are made from modacrylic fibers that are designed to look and feel like real hair.

⇨ 인조모는 모다아크릴릭 섬유에서 부터 만들어진 것으로 진짜 인모와 비슷하게 느껴지고 보여지도록 만든 것이다.

♠ The fibers have the ability to return to their originally styled shape by lightly spraying with water and drying from underneath.

⇨ 이러한 섬유들은 가볍게 물을 뿌리고 아래쪽에서 부터 건조시키면 처음의 스타일도 돌아오는 능력을 가지고 있다.

♠ In addition, synthetic fiber doesn't absorb odor and doesn't react to climate changes as readily as natural hair wigs.

⇨ 게다가, 인조섬유는 악취를 흡수하지 않고 인모가발처럼 기후에 따른 변화가 별로 없다.

Dialogue

i n t r o d u c e

PART 01 | 인사

hairdesser : Good morning, ma'am

고객님, 어서오세요.

customer : Good morning

안녕하세요.

hairdesser : Have you been to our salon before?

그전에 저희 샵을 방문하신 적이 있나요?

customer: No, this is the first time

아니요. 이번이 처음입니다.

r e c e p t i o n

PART 02 | 리셉션

hairdesser : What can we help you?

무엇을 도와드릴까요

customer : I wanna get a haircut.

머리를 자를까 하구요.

hairdesser : Did you have a reservation?

예약은 하셨나요?

customer : Yes, I did.

네. 그렇습니다.

hairdesser : Ok. Can I have a name?

네. 성함이 어떻게 되시나요

customer : I'm Nadia

나디아라고 합니다.

hairdesser : I will keep your belongings.

제가 물건들을 보관해드리겠습니다.

Take off your coat and put on this gown.

코트를 벗으시고 이 가운을 입으세요.

customer : Thank you.

고맙습니다.

hairdesser : You are welcome. Step this way, please.

천만에요. 이쪽으로 오세요

hair shampoo

PART 03 | 웨이팅

hairdesser : I'm sorry, but you have to wait.
죄송하지만, 기다리셔야 할 것 같아요.
It will take about 15 minutes.
15분 정도 걸릴겁니다.

customer: Ok. no ploblem.
네. 괜찮습니다.

hairdesser : Thank you. Shall I bring to the magazine?
감사합니다. 잡지를 가져다 드릴까요?

customer : Cool, Thank you.
네. 감사합니다.

hairdesser : Would you like something to drink?
차 좀 드시겠어요?

customer : That would be nice. Can I have a cup of coffee without sugar?
아 네 좋아요. 설탕 안 넣은 커피 한잔 마실 수 있을까요?

hairdesser : Sure. Wait a minute, plese
물론이죠. 잠시만 기다려 주세요.

customer : Thank you.
고맙습니다.

hairdesser : You are welcome.
천만해요.

PART 04 | 헤어커트

hairdesser : How would you like your hair today?

오늘 어떤 스타일을 하고 싶으신가요

customer : I wanna get a haircut.

머리를 자를까 하구요.

hairdesser : Ok. How short do you want your hair?

네. 얼마나 짧게 하시고 싶으세요?

customer: I wanna just trim the ends.

저는 그냥 끝에만 살짝 다듬고 싶어요.

And I'd like to have a soft fringe.

그리고 가벼운 앞머리도 있으면 좋겠어요.

hairdesser : Ok then. I will just give you a trim and tidy up today.

네, 그럼 오늘은 살짝치고 다듬어 드릴께요.

h a i r   s h a m p o o

PART 05 | 헤어샴푸

hairdesser : Take a seat here plese.

여기에 앉으세요.

customer : Thanks

고마워요.

hairdesser : Lean your neck against the sink plese.

목을 뒤로 기대세요.

customer : Ok.

네.

hairdesser : Is the temperature of the water right for you?

물의 온도가 괜찮으세요?

customer: Well. It is a bit cold for me.

음. 약간 차가운것 같아요.

hairdesser : Ok. How about now?

네. 지금은 어떠세요?

customer : It is good.

네. 좋아요.

hairdesser : Ok. We are done. Sit up plese.

자. 다 되셨습니다. 앉으세요.

customer : Thank you.

감사합니다.

coloring

GN-20
GN-30
GN-40
GN-50
GN-60
GN-70
GN-80
GN-90
GN-100
GR-30
GR-40
GR-50
GR-60
GR-70
GR-80
R-90
R-100
GY-30
GY-40
GY-50
GY-60
GY-70
GY-80
Y-90
Y-100
GM-50
GM-60
GM-70
GM-80
GM-90
GM-100
GAB-7
AB-9
LIGHTER
YELLOW-9
ORANGE-7
RED-7
ASH-9
VIOLET-4
GREEN-3
BLUE-3

PART 06 | 컬러링

hairdesser : What color do you have in mind?

무슨 색으로 하실까 생각하시나요?

customer : I wanna go for a light color.

좀 밝은 색으로 하고 싶어요.

hairdesser : Do you wanna get the ash blonde?

차가운 느낌의 금발은 어떠세요?

customer : That sounds nice.

괜찮을 것 같으네요.

Do I need to bleach it?

탈색해야 되는거죠?

hairdesser : Yes. Your hair is dark at the moment.

네. 지금 머리색이 어두우시니까요.

customer: I see.

네 알겠어요.

hairdesser : I think your fair skin would go on well with bright colors.

제 생각에 하얀피부라서 밝은 색이 잘 어울리실 것 같으네요.

customer : Oh, Thanks.

아, 고맙습니다.

hairdesser : Then I will start bleaching. It will take about an hour.

그럼 탈색부터 시작할께요. 한시간 정도 걸릴꺼에요.

customer: Ok. good.

아 네 좋아요.

p e r m a n e n t   w a v e

PART 07 | 퍼머넌트 웨이브

hairdesser : What style would you like?
어떤 스타일을 원하세요?

customer: I want more volume in my hair.
머리에 볼륨이 더 있었으면 해서요.
Actually, I want to do the loose curls.
사실은. 좀 굵은 컬을 하고 싶어요.

hairdesser : Sure. I can do that.
그럼요. 그렇게 해드릴 수 있어요.

customer: Perfect
네.

PART 08 | 블로우 드라이

hairdesser : Hello. How are you?
안녕하세요. 오늘 어떠세요?
How can I help you?
무엇을 도와드릴까요?

customer : Fine. Thank you.
네. 좋아요.
I need to have my hair blow dried.
머리 드라이를 해야 할 것 같아서요.
I got invited to the great party tonight.
오늘밤 훌륭한 파티에 초대를 받아서요.

hairdesser : Brilliant. A proper blow dry can add fullness and volume to your hair.
멋지네요. 적절한 드라이는 모발에 풍성함과 볼륨감을 드릴꺼에요.

h a i r t r e a t m e n t

PART 09 | 헤어트리트먼트

hairdesser : Hi, How can I help you?

안녕하세요 .무엇을 도와드릴까요?

customer : My hair is tangled and dry.

제 머리가 엉키고 건조합니다.

hairdesser : Our special hair treatment will help your hair more smooth, shiny and manageable.

저희의 특별한 헤어트리트먼트는 모발을 더욱 부드럽고윤기 있으며 관리하기 쉽게 해드릴거에요.

customer : Really? That's brilliant.

정말요? 너무 좋은데요.

hairdesser : Yes. We have a very good range of hair care products.

네. 저희는 정말 좋은 헤어케어 제품을 가지고 있습니다.

customer : Ok. I will try it.

좋아요. 한번 해볼께요.

hairdesser : Good. They are all made with natural ingredients.

네 . 모두 천연 제품들로 만들어진 제품들이구요.

They are all gentle and nourishing.

자극이 적고 영양분을 공급해주는 제품들입니다.

You should be able to achieve a smooth and silky result.

사용 후에 부드럽고 매끈한 결과를 얻을 수 있을 꺼예요.

customer : I see. Thank you

네 알겠어요. 고마워요.

f i g u r e s

PART 10 | 계산

hairdesser : We are done.

시술 다 마치셨습니다.

customer : Ok. How much is it?

네. 얼마인가요?

hairdesser : It is forty thousands won.

사만원입니다.

customer : Ok. Do you take credit cards?

네. 신용카드 받으시죠?

hairdesser : Yes, We do.

네. 그렇습니다.

customer : Ok, here it is.

네. 여기있어요.

hairdesser : Would you sign here plese?

여기에 사인해 주시겠어요?

customer : sure.

네.

hairdesser : Thank you very much.

대단히 감사합니다.

s e e   g u e s t s   o u t

PART 11 | 배웅

hairdesser : Did you leave any of your belongins with us?

남기신 물건 있으신가요?

customer : I left my coat and a handbag.

코트랑 핸드백을 맡겼었어요.

hairdesser : Ok. Are these yours?

네. 이게 손님건가요?

customer : Yes. They are. Thanks.

네. 제 꺼에요. 감사합니다.

hairdesser : Thanks for coming. We will see you next time.

와주셔서 감사합니다. 다음에 또 뵐께요.

customer : Thank you. Bye.

감사합니다. 안녕히 계세요.

r e s e r v a t i o n

PART 12 | 예약

hairdesser : Hello. This is the Nadia beauty salon, Kate speaking.

안녕하세요. 나디아 뷰티살롱의 케이트입니다.

How can I help you?

무엇을 도와드릴까요?

customer : Hello. I'd like to make an appointment for coloring.

안녕하세요. 염색하는거 예약하려구요.

hairdesser : Ok. When would you like to an appointment?

네. 예약은 언제로 해드릴까요?

customer : This friday plese.

이번주 금요일날로 해주세요.

hairdesser : Let me see. Ok. You can make it for friday.

잠깐만요. 네 이번주 금요일날로 예약해 드리겠습니다.

And What time would you like to have?

몇시로 잡아드릴까요?

customer : Can I make it 2?

오후 2시쯤 가능할까요?

hairdesser : Ok. And Can I take your name and contact number please?

네. 성함이랑 연락 가능한 번호를 알려주시겠어요?

customer : Sure. My name is rozina and my cell phone is 010-5555-1234

네. 제 이름은 로지나이고 전화번호는 010-5555-1234에요.

hairdesser : Ok. Your booking for friday is done.

네. 금요일 예약이 다 되셨습니다.

Home care

PART 13 | 홈케어

customer : How can I protect my hair?

제 머리를 어떻게 보호하죠?

hairdesser : Use moisturizing shampoo and don't rub too harsh after shampoo.

보습샴푸를 쓰고 샴푸 후에는 너무 거칠게 문지르지 마세요.

Wrap the hair with a towel and squeeze and rub gently.

타월로 모발을 싸고 물기를 눌러준 뒤 조심스럽게 문질러 닦으세요.

customer : I see. Thank you for the tip

아 네. 조언 고마워요.

저자 약력

김세욱 안산대학교 뷰티아트과 학과장
김유정 여주대학교 뷰티디자인과 학과장
김영주 오산대학교 피부미용과 학과장

미용영어

2013년 3월 1일 초판 인쇄
2013년 3월 5일 초판 발행

저 자 김세욱 김유정 김영주
발 행 인 조규백
발 행 처 도서출판 구민사
(150-034) 서울특별시 영등포구 영등포4가 104-1번지 동서빌딩 604호
전 화 02.701.7421~2
팩 스 02.3273.9642
홈페이지 www.kuhminsa.co.kr
등 록 제14-29호(1980년 2월 4일)
I S B N 978-89-7074-745-3 93150
값 19,000원